中国可再生能源发展报告
2023 年度简本

水电水利规划设计总院　编

·北京·

图书在版编目（CIP）数据

中国可再生能源发展报告. 2023年度简本 / 水电水利规划设计总院编. -- 北京 : 中国水利水电出版社, 2024. 7. -- ISBN 978-7-5226-2625-3

Ⅰ. F426.2

中国国家版本馆CIP数据核字第2024KB9571号

责任编辑：张晓

书　　名	**中国可再生能源发展报告 2023 年度简本** ZHONGGUO KEZAISHENG NENGYUAN FAZHAN BAOGAO 2023 NIANDU JIANBEN
作　　者	水电水利规划设计总院　编
出版发行	中国水利水电出版社 （北京市海淀区玉渊潭南路1号D座　100038） 网址：www.waterpub.com.cn E-mail：sales@mwr.gov.cn 电话：（010）68545888（营销中心）
经　　售	北京科水图书销售有限公司 电话：（010）68545874、63202643 全国各地新华书店和相关出版物销售网点
排　　版	中国水利水电出版社微机排版中心
印　　刷	北京科信印刷有限公司
规　　格	178mm×248mm　16开本　2.75印张　50千字
版　　次	2024年7月第1版　2024年7月第1次印刷
定　　价	58.00 元

凡购买我社图书，如有缺页、倒页、脱页的，本社营销中心负责调换

版权所有·侵权必究

编 委 会

主　　任	李　昇　易跃春
副 主 任	王忠耀　何　忠　赵全胜　余　波　王富强　黎扬佳
	郭建欣　薛联芳　钱钢粮　李光顺
主　　编	赵增海　张益国　彭才德
副 主 编	谢宏文　郭雁珩　姜　昊　彭烁君　辛颂旭　朱方亮
	薛建峰　喻葭临　李修树
编写人员	徐志成　谢越韬　刘玉颖　周　力　常昊天　肖段龙
	司俊龙　陈　长　杜斯佳　刘双希　霍晶莹　李少彦
	弭　辙　武明鑫　许　月　任伟楠　赵多苍　王　欢
	王　敏　李彦洁　薛美娟　周小溪　王伶俐　吕　嵩
	李晓曦　韩　冬　艾　琳　陈国生　柯　文　程　立
	胡耀飞　傅兆庆　耿大洲　贾浩帅　于雄飞　郭　豹
	刘春高　孟　栩　段　聪　李宏宇　郝　宇　张步恩
	王昊轶　赵国斌　郭瑾瑜　邱　辰　乔　勇　崔正辉
	谢　豪　郑　静　魏国强　宗万波　仇雅静　孙漪蓉
	刘　超　韦惠肖　田子婵　何　伟　郝军刚　查　浩

目录

1 发展现状：中国成为全球可再生能源发展领头羊　　01

2 开发进展：各类可再生能源建设全面提速　　05

3 利用情况：可再生能源发电量和利用效率持续提升　　08

4 技术发展：技术持续创新推动产业进步　　11

5 展望未来：中国可再生能源将成为全球构建清洁低碳未来的关键力量　　14

1 发展现状

中国成为全球可再生能源发展领头羊

中国可再生能源发电累计装机容量突破 1500GW 大关

截至 2023 年年底,中国可再生能源发电装机容量达 1516.4GW,同比增长 24.9%,占全国发电总装机容量的 51.9%,在全球可再生能源发电总装机中的占比接近 40%。在可再生能源发电装机中,常规水电装机容量 370.6GW(占比 12.7%)、抽水蓄能装机容量 50.9GW(占比 1.7%)、风电装机

容量441.3GW（占比15.1%）、太阳能发电装机容量609.5GW（占比20.9%）、生物质发电装机容量44.1GW（占比1.5%）。太阳能发电和风电总装机容量超过1000GW，从装机上分别跃升为中国第二、第三大电源。可再生能源在中国能源结构中的重要性日益凸显，人均可再生能源装机容量超过1kW。2023年中国各类电源装机容量及占比如图1.1所示。

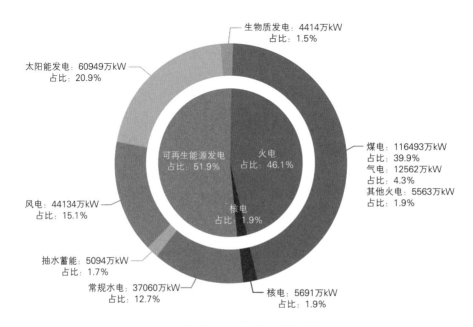

图1.1　2023年中国各类电源装机容量及占比

2023年中国可再生能源发电新增装机容量303GW

自2020年以来，中国可再生能源发电连续三年新增装机容量超过100GW。2023年中国可再生能源发电新增装机容量增长至303GW，超过世界其他国家的总和，占中国新增电力装机容量的84.9%，约占全球可再生能源发电新增装机容量的59.8%。其中，常规水电新增装机容量2.43GW，抽水蓄能新增装机容量5.15GW；风电新增装机容量75.66GW，光伏发电新增装机容量216.88GW，生物质发电新增装机容量2.82GW。

中国可再生能源发电量达 2951TW·h

2023 年，中国可再生能源发电量 2951TW·h，占全社会用电量的 32%。其中，水电、风电、太阳能发电、生物质发电量分别为 1284TW·h、886TW·h、583TW·h 和 198TW·h，占中国全口径发电量的比例分别为 13.8%、9.5%、6.3% 和 2.1%；风电、光伏发电量占全社会用电量的比例超过 15%，同比增长 24%，成为拉动非化石能源消费占比提升的主力。2023 年中国各类电源年发电量及占比如图 1.2 所示。

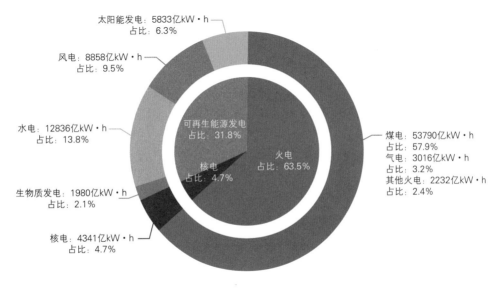

图 1.2　2023 年中国各类电源年发电量及占比

中国生物质能非电利用及其他可再生能源利用稳步推进

2023 年，中国生物质能非电利用及地热等其他可再生能源利用规模持续扩大。生物质能非电利用量折合标准煤约 2098 万 t，同比增长 24.4%，生物天然气、生物质固体成型燃料、燃料乙醇和生物柴油年产量增幅明显。地热能规模化开发格局初步形成，地热能开发利用以供暖（制冷）为主，浅层地热能供暖项目规模化开发主要集中在华北地区和长江中下游地区，中深层地热能供暖项目的扩大得益于北方地区冬季清洁取暖政策的推动。

中国可再生能源政策体系持续推进

中国在法制化、市场化、技术创新、碳减排、绿证制度、标准体系建设等方面出台多项政策，以促进可再生能源健康发展。《国土空间规划法》《中华人民共和国能源法》和《中华人民共和国可再生能源法》（修改）被列为重要立法项目。市场化运营通过《电力现货市场基本规则（试行）》等文件得到规范，新能源市场化交易扩大。技术创新方面，发布了《关于推动能源电子产业发展的指导意见》（工信部联电子〔2022〕181号），探索智慧能源系统。碳减排领域，推进温室气体自愿减排，发布相关管理办法和项目方法学。绿证制度通过《关于做好可再生能源绿色电力证书全覆盖工作 促进可再生能源电力消费的通知》（发改能源〔2023〕1044号）得到加强。《碳达峰碳中和标准体系建设指南》为风力发电、光伏发电等领域提供标准。此外，中国针对光伏用地、光热发电、风电场改造等方面也颁布了政策，促进新能源行业健康发展。

中国可再生能源产业成为全球清洁能源转型的重要推动力

中国在推动国内清洁能源发展的同时，积极贡献技术和解决方案，助力全球能源转型。中国可再生能源产业为全球200多个国家和地区提供高质量产品和服务，降低全球成本。技术创新方面，中国在异质结电池、钙钛矿-晶硅叠层光伏电池等领域取得显著进展，储能技术和智能电网技术也实现突破。2023年，中国能源转型投资额达到6760亿美元❶，占全球投资的38%，位居首位，这些投资促进了国内外能源结构的转型，支持全球清洁能源项目，推动构建可持续和环境友好的能源体系。

---

❶ 数据来源为彭博新能源财经（BNEF）的《2024年能源转型投资趋势》（*Energy Transition Investment Trends 2024*）报告。

2 开发进展

各类可再生能源建设全面提速

中国常规水电已建装机容量370.6GW，开发向高海拔地区推进

截至2023年年底，中国常规水电装机容量达到370.6GW，主要集中在西南、华中、华南、华东和西北区域。2023年中国常规水电新增装机容量2.43GW；在建大型常规水电站合计装机容量约30.64GW；新核准的大型常规水电装机容量约4.15GW，预计年发电量约18600GW·h。未来，水电开发将更多地向流域上游高海拔地区推进。

中国抽水蓄能总投产规模达到 51GW，核准在建规模达 179GW

截至 2023 年年底，中国抽水蓄能电站总装机规模达到 51GW，其中华东、华北和南方为重点发展区域，西北区域电网首次实现投产。2023 年中国抽水蓄能新增投产规模 5.15GW，新增核准在建的抽水蓄能总装机规模 179GW，其中华中、华东、西北区域电网核准在建规模均超过 30GW。

中国风电装机保持高速增长，陆上风电趋向大基地集中，海上风电建设呈基地集群态势，机组大型化趋势明显

截至 2023 年年底，中国风电装机容量累计达 441.3GW，占全国全口径发电总装机容量的 15.1%，占全球风电总装机容量的 43%。其中，陆上风电累计装机容量 404GW，"三北"（东北、华北、西北）地区风电累计装机容量占全国风电总装机容量的 66%；海上风电累计并网规模 37.3GW，连续三年全球第一。2023 年，中国风电新增装机容量 75.6GW，创历史新高，陆上风电和海上风电分别新增装机容量 69.3GW 和 6.3GW。

中国太阳能发电装机规模增长再创历史新高，集中式光伏电站加速建设，"沙戈荒"基地项目助力，分布式光伏发电稳步发展，光热发电规模化发展快速推进

截至 2023 年年底，中国太阳能发电装机容量 609.5GW，占全国电源总装机容量的 20.9%，成为全国第二大电源。光伏发电新增装机容量 216.9GW，占全球新增光伏发电装机容量的 62.5%；累计装机容量 608.9GW，全球领先。其中，集中式光伏发电累计装机容量 355.5GW，新增装机容量 121.1GW；分布式光伏发电累计装机容量 253.4GW，新增装机容量 95.8GW。光热发电并网装机容量 0.57GW。

生物质发电装机规模持续增长，生活垃圾焚烧发电项目进入稳定发展期，农林生物质发电项目建设规模平稳

截至 2023 年年底，中国生物质发电利用量折合标准煤约 5940 万 t，占生物质总开发利用量的 73.9%；生物质发电累计并网装机容量 44.1GW，同比增长 6.8%，新增装机容量 2.8GW，其中农林生物质发电新增装机规模 0.6GW，占比 23%；生活垃圾焚烧发电新增装机规模 1.9GW，占比 68%；沼气发电新增装机规模 0.3GW，占比 9%。

中国新型储能累计装机规模超 31GW，锂离子电池储能占比 97.4%

截至 2023 年年底，中国新型储能累计装机规模超 31GW，近三年增长了近 10 倍，华北地区和西北地区累计装机规模占比超过全国新型储能总装机规模的 50%。锂离子电池储能占比 97.4%，压缩空气储能进入示范推广阶段。

中国地热能规模化开发格局初步形成，地热开发力度逐步增强

中国地热能主要用于供暖（制冷），次为温泉理疗和农业养殖，少量用于发电。华北地区和长江中下游地区浅层地热能供暖规模化开发，北方地区中深层地热能供暖项目在河北、河南、山东、天津、陕西、山西等六省（直辖市）规模扩大。

中国可再生能源制氢发展步伐加快

中国氢气生产和需求居全球领先地位，2023 年总产量达 4291 万 t。可再生能源制氢产能 7.8 万 t/年，在建项目产能约 80 万 t/年，已备案项目产能超 600 万 t/年。"三北"地区是主要产能聚集地。

中国海洋能开发取得突破

2023 年中国海洋能行业取得重大突破，从理论研究和小型试验转向大型工程样机示范。世界首台兆瓦级漂浮振荡体式波浪能发电装置"南鲲"号在广东珠海试运行，中国首台超 100kW 振荡水柱式波浪发电装备"华清号"开工建造，20kW 海洋漂浮式温差能发电装置在南海完成海试。

3 利用情况

可再生能源发电量和利用效率持续提升

中国水电年发电量 1283.6TW·h，水电年平均利用小时数 3133h

2023 年，中国水电年发电量 1283.6TW·h，占全部电源年总发电量的 13.8%。中国水电年平均利用小时数 3133h，中国有效水能利用率达 99.0%，创 2016 年以来新高。2014—2023 年中国水电年发电量及占比变化趋势如图 3.1 所示。

图 3.1 2014—2023 年中国水电年发电量及占比变化趋势

风电发电量在电力供应结构中比重持续提升，年利用率达到 97.3%

2023 年，中国风电年发电量达到 885.8TW·h，占全部电源年总发电量的 9.5%。沿海省份积极发展海上风电，作为绿色能源转型的重要措施。中国风电年平均利用小时数 2225h，风电年利用率达到 97.3%。2014—2023 年中国风电累计并网装机容量及年利用率变化趋势如图 3.2 所示。

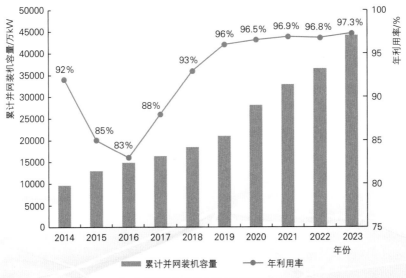

图 3.2 2014—2023 年中国风电累计并网装机容量及年利用率变化趋势

光伏发电装机快速增长的同时利用率保持较高水平，实现了98％的年平均利用率

截至2023年年底，中国太阳能发电量达到583.3TW·h，占全口径发电量的6.3％。光伏发电量582.3TW·h，其中集中式光伏发电量359.1TW·h；分布式光伏发电量223.2TW·h，在华北、华东和华中地区发展迅速，三地区分布式光伏发电量约占分布式光伏总发电量的87.1％。光热发电量1TW·h。2023年光伏发电实现了98％的年利用率，与2022年基本持平。2016—2023年中国光伏发电累计并网装机容量及年利用率变化趋势如图3.3所示。

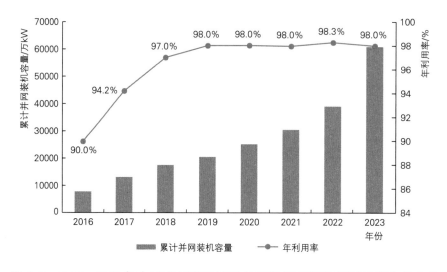

图3.3 2016—2023年中国光伏发电累计并网装机容量及年利用率变化趋势

中国生物质发电量保持平稳增长，年发电小时数保持稳定

2023年，中国生物质发电年发电量198TW·h，占全部可再生能源总发电量的6.7％。其中，农林生物质发电年发电量55TW·h，同比增长6.4％；生活垃圾焚烧发电年发电量139.4TW·h，同比增长9.9％；沼气发电年发电量3.6TW·h，同比降低10.0％。生物质发电年平均利用小时数4626h，较2022年增加54h。

4 技术发展

技术持续创新推动产业进步

2023年，中国通过技术创新推动可再生能源产业快速发展。水电工程在多个领域取得技术突破，风电装备和光伏产业规模持续扩大，光热技术达到国际先进水平，生物质发电和新型储能技术不断进步。同时，中国在抽水蓄能、地热能、氢能和海洋能技术上也取得显著进展。

常规水电和抽水蓄能：先进的勘测技术和装备为复杂建设条件下水电工程提供了坚实的技术支持

先进的勘测技术和装备为中国水电工程提供了技术支持，

如 1000m 级垂直钻孔技术和 800m 级、1000m 级超深水平钻探技术。青藏高原地区水电工程通过创新技术解决了高寒、高海拔等技术难点。技术创新和研发投入推动了水电机组技术提升和新型设备标准化。智能化建设创新应用实现了工程全生命周期数字化管控，提高了抽水蓄能电站群运行管理的集约化、智能化水平。

风电：风电机组向大型化方向持续推进，关键部件技术不断突破，产业体系持续健全

2023 年，中国风电机组持续大型化，陆上风电平均单机容量 5.5MW，最大 11MW；海上风电平均单机容量 9.5MW，最大 20MW。中国风电机组关键部件技术不断突破，风电勘测—设计—施工—运维产业体系持续健全，填补了海上风电运维领域空白。

太阳能发电：光伏电池转换效率持续提升壮大光伏产业规模

2023 年，中国光伏电池市场技术路线结构转变，N 型电池片占比提升至 26.5%。晶硅电池转化效率提升，P 型 PERC 电池平均转换效率为 23.4%，N 型 TOPCon 和异质结电池平均转换效率分别达到 25.0% 和 25.2%。薄膜电池技术升级，碲化镉（CdTe）和铜铟镓硒（CIGS）电池平均转换效率分别提升至 15.8% 和 16.5%。组件功率提升，双面组件市场占比 67%，半片组件占比 97.1%。光热技术能力提升，产业链完整，光热电站单体规模向 200MW 发展。

生物质能：生物质发电和供热技术水平不断提升

中国生物质发电和供热技术持续进步，高温高压锅炉广泛应用，锅炉寿命延长。发电机组效率提高至 37%，生物质热电联产技术在多领域规模化应用。生物质能综合利用技术不断创新，生物质制氢和绿氢耦合生物质制甲醇技术取得进展，中国成功将生物航煤技术商业化。

新型储能：锂离子电芯单体规模和技术性能进一步提升

中国锂离子电芯技术性能提升，300Ah+ 电池替代 280Ah 产品，企业布局

更高容量电池。磷酸铁锂电芯能量密度达 160Wh/kg，转换效率 85%，循环寿命 6000 次以上。安全性能提升，液冷技术取代风冷。构网型储能技术取得进展，压缩空气储能技术取得突破。

地热能：深部地热钻井和长距离水平钻井技术取得突破

中国在深地热钻探和水平钻井技术上取得突破，钻探技术深入至 5000m，实现千米水平直井对接。地热能发电装备和试验创新，突破 20MW 双工质地热发电装备。

氢能：电解槽大型化、高效化发展持续推进

中国在电解槽技术方面取得进展，碱性电解槽最大制氢量 3000Nm^3/h，PEM 电解槽最大单槽产氢量达 400Nm^3/h。储氢技术发展，Ⅳ型储氢瓶质量储氢密度超 6wt%，镁基固态储氢产业化，液态储氢技术准备示范应用。管道输氢技术实现规模化，火电掺氢（氨）技术得到验证。

海洋能：兆瓦级漂浮式波浪能发电平台、温差能发电透平等技术取得突破

中国海洋能技术在多个领域取得显著进展，特别是在兆瓦级漂浮式波浪能发电平台、后弯管式波浪能发电机理、温差能发电透平等技术方向。其中，波浪能发电系统实现了从百千瓦级到兆瓦级的跨越；波浪能发电船开创了新的技术路线，建立了具有自主知识产权的核心技术体系。20kW 海洋漂浮式温差能发电装置完成了深海试验。

5　展望未来

中国可再生能源将成为全球构建清洁低碳未来的关键力量

预计 2024 年，中国在可再生能源领域的投产规模将显著增长。常规水电投产规模预计达 6GW，大型常规水电核准规模约 3GW。抽水蓄能发电投产规模预计达到 6GW，运行总规模将增至 57GW，核准规模约 40GW。风电增长迅速，预计并网装机容量约 70GW。太阳能发电并网容量预计约 190GW。这些发展将推动中国能源结构的绿色转型。

中国可再生能源发展取得了举世瞩目的成就。未来，中国将继续坚持创新驱动，推动可再生能源高质量发展，为实现碳达峰碳中和目标，构建清洁低碳、安全高效的能源体系作出更大贡献。

声　　明

　　本报告内容未经许可，任何单位和个人不得以任何形式复制、转载。

　　本报告相关内容、数据及观点仅供参考，不构成投资等决策依据，水电水利规划设计总院不对因使用本报告内容导致的损失承担任何责任。

　　如无特别注明，本报告各项中国统计数据不包含香港特别行政区、澳门特别行政区和台湾省的数据。 部分数据因四舍五入的原因，存在总计与分项合计不等的情况。

　　本报告部分数据及图片引自国际可再生能源署（International Renewable Energy Agency）、国际水电协会（International Hydropower Association）、国家统计局、国家能源局、中国电力企业联合会等单位发布的数据，以及 *Renewable Capacity Statistics* 2024、2023 年全国电力工业统计数据、中国光伏产业发展路线图（2023—2024 年）、中国太阳能热发电行业蓝皮书 2023、*Energy Transition Investment Trends* 2024 等统计数据报告，在此一并致谢！

Disclaimer

This report shall not be reproduced or disseminated in any form without permission.

The content, data, and viewpoints presented in this report are for reference only and should not be construed as investment advice or advisory opinions for other decisions. Hence, the China Renewable Energy Engineering Institute (CREEI) disclaims any liability for the losses arising from use of this report and/or its contents.

Unless otherwise noted, statistics of China in this report do not include that of the Hong Kong Special Administrative Region, the Macao Special Administrative Region, and Taiwan Province. Due to rounding, sub totals of some items may not equal the overall sum.

Part of the data and images in this report are derived from the data released by the International Renewable Energy Agency (IRENA), the International Hydropower Association (IHA), the National Bureau of Statistics of China, the National Energy Administration of China and the China Electricity Council. Other reference materials include *Renewable Capacity Statistics* 2024, 2023 *National Electricity Industry Statistics*, *China PV Industry Development Roadmap* (2023 – 2024), *Blue Book of China's Concentrating Solar Power Industry* 2023, and *Energy Transition Investment Trends* 2024. We hereby extend our acknowledgement to all these sources.

The renewable energy development in China has achieved remarkable accomplishments. In the future, China will adhere to the innovation-driven principle, continuously promote the high-quality development of renewable energy, and make greater contributions to realizing the goals of carbon peaking and carbon neutrality, and building a clean, low-carbon, safe and efficient energy system.

5 Looking forward

China's renewable energy will become the key force for a clean and low-carbon shared future

It is estimated that by the end of 2024, the renewable energy power generation capacity will have increased significantly. The expected added capacity of conventional hydropower will be 6 GW, with about 3 GW approved. The newly added capacity of pumped storage is expected to reach 6 GW, making the total capacity of 57 GW, while another 40 GW is expected to be approved for construction. It is expected that the capacity additions for wind power and solar power are about 70 GW and 190 GW respectively. The above-mentioned expected development will further promote the China's green energy transition.

innovated, and advances have been made in 20 MW dual-working medium thermal power generation equipment.

Hydrogen: electrolyzers continued to scale up with high efficiency

China has made progress in electrolyzer technology. The maximum hydrogen production of alkaline electrolyzer per unit was 3,000 Nm^3/h, and that of PEM electrolyzer pre unit was 400 Nm^3/h. With the development of hydrogen storage technology, the mass hydrogen storage density of IV-type hydrogen cylinder exceeded 6 wt%, magnesium-based solid hydrogen storage was industrialized, and liquid hydrogen storage technology was prepared for demonstration application. The technologies of hydrogen transportation by pipelines were available for application on a large scale, and the technology of thermal power generation with doped-hydrogen (doped-ammonia) has been verified.

Marine energy: breakthroughs were achieved in megawatt-scale floating wave energy generation platforms, ocean thermoelectric power generation turbines, and other technologies

China has made remarkable progress in many fields of marine energy technologies, especially in the technologies of megawatt-scale floating wave energy generation platforms, backward-bent duct buoy (BBDB) wave energy generation mechanism and ocean thermoelectric power generation turbines. Among them, the wave power generation system realized a leap from 100 kW to 1 MW; a new technical route was created for wave power generation ship and a core technical system with independent intellectual property rights was established. The 20 kW offshore floating thermoelectric power generation turbines completed the deep-sea test.

China Renewable Energy Development Report 2023 Abridged Version 17

respectively. With the upgrading of thin-film cell technology, the efficiency of CdTe and CIGS cells was increased to 15.8% and 16.5% respectively. The capacity of modules increased, with the market share of double-sided modules accounting for 67% and half-cell modules accounting for 97.1%. The CSP technology was further improved, with a complete industrial chain in place. The increased capacity of CSP plant per unit is progressing towards 200 MW.

Biomass energy: the technology of biomass power generation and heating has been advanced constantly

With the continuous upgrading of biomass power generation and heating technologies in China, high-temperature and high-pressure boilers have been widely used, and the lifespan of boilers was prolonged. The efficiency of generator units has been increased to 37%, and biomass cogeneration technology was applied to many fields on a large scale. The integrated utilization technologies of biomass energy were improved, and progress has been made in the technologies of hydrogen production from biomass and methanol production from biomass combined with green hydrogen. Bio-jet fuel technology was successfully commercialized in China.

New energy storage: the unit scale and technical performance of lithium-ion cells were further improved

The technical performance of lithium-ion cells in China was improved, with replacement of 280 Ah cells by 300 Ah + cells, and enterprises were strategizing for larger capacity batteries. The lithium-iron phosphate cells featured an energy density of 160 Wh/kg, with a conversion efficiency of 85% and a cycle life of more than 6,000 times. Safety performance was improved, with liquid-cooling technology replacing air cooling technology. Progress had been made in the grid-forming control energy storage technology, and breakthroughs were made in the compressed air energy storage technologies.

Geothermal energy: breakthroughs have been made in deep geothermal drilling and extended-reach horizontal drilling technologies

Breakthroughs were made in deep geothermal drilling and horizontal drilling technologies in China, with the realization of the drilling distance up to 5 km and the connection of 1 km horizontal and vertical wells. Geothermal power generation equipment and tests were

Conventional hydropower and pumped storage: advanced surveying technologies and equipment provided solid technical support for hydropower projects under complex construction conditions

Advanced surveying technologies and equipment provided technical support for hydropower projects, such as 1km-level vertical drilling technology, 800m-level and 1km-level ultra-deep horizontal drilling technologies. Technical difficulties of hydropower projects in the Qinghai-Xizang Plateau, such as the frigidity and other high-altitude problems, were resolved through innovative technologies. Technological innovation and Research & Development (R&D) investment promoted the technical upgrading of hydropower operational units and the standardization of new equipment. The innovative application of intelligent construction realized the digital management and control of the full-life cycle of projects, and improved the intensification and intellectualization levels of the operation and management of pumped storage power plants.

Wind power generation: wind turbine generators continued to advance towards larger sizes, achieving continuous breakthroughs in key component technologies, with the industrial system being constantly improved

In 2023, the wind turbines in China continued to advance towards larger sizes. The average power generation capacity per unit of onshore wind turbine was 5.5 MW, with a maximum of 11 MW. The average unit capacity of offshore wind turbine was 9.5 MW, with a maximum of 20 MW. Continuous breakthroughs were made in technologies of key components of wind turbines in China, and the industrial system of surveying - design - construction - operation and maintenance has been improved constantly for wind power plants, filling the gaps in operation and maintenance of offshore wind projects.

Solar power generation: the conversion efficiency of solar PV cells has been enhanced continuously, scaling up solar PV industry

In 2023, the technical routes of solar PV cells industry in China changed structurally, and the share of N-type cells was increased to 26.5%. The conversion efficiency of crystalline silicon cells was enhanced. The average conversion efficiency of P-type PERC cells was 23.4%, and that of N-type TOPCon and heterojunction cells was 25.0% and 25.2%

4 Technological development

Continuous technological innovation promoted industrial progress

In 2023, China promoted the rapid development of renewable energy industry through technological innovation. Technological breakthroughs were observed in many fields of hydropower engineering, the scale of wind power equipment and solar PV industry continued to expand, technology of CSP reached the world-class level, and biomass power generation and new energy storage technologies has been continuously advancing. Meanwhile, China has also attained remarkable achievements in pumped storage, geothermal energy, hydrogen production and marine energy technologies.

The biomass power generation in China maintained steady growth, and the annual power generation hours kept stable

In 2023, the annual biomass power generation in China was 198 TWh, accounting for 6.7% of the total renewable power generation. The agricultural and forestry biomass power generation was 55 TWh with a year-on-year increase of 6.4%; domestic waste incineration power generation was 139.4 TWh with a year-on-year increase of 9.9%; the biogas power generation was 3.6 TWh with a decrease of 10.0%. The annual average utilization hours of biomass power generation reached 4,626 h with an increase of 54 h compared to 2022.

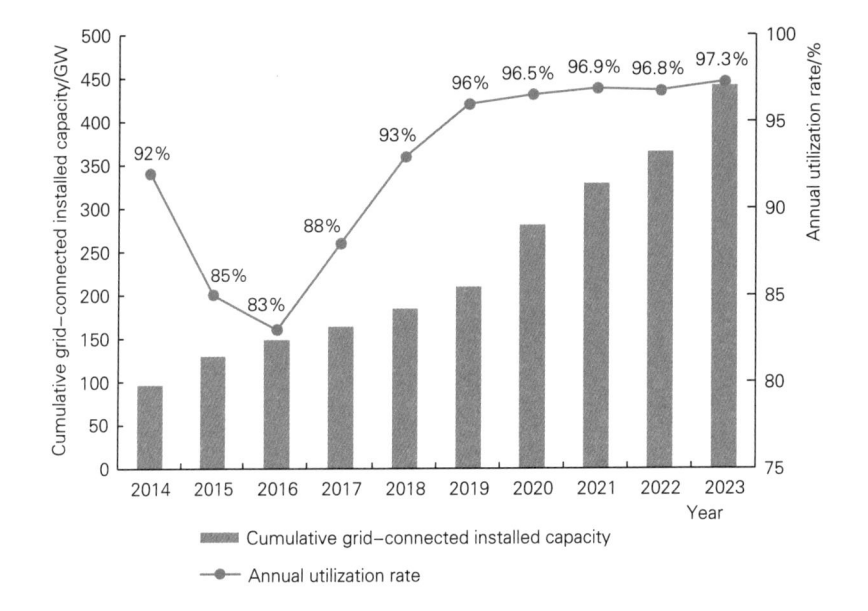

Figure 3.2　Cumulative grid-connected installed capacity and annual utilization rate of wind power in China, 2014 – 2023

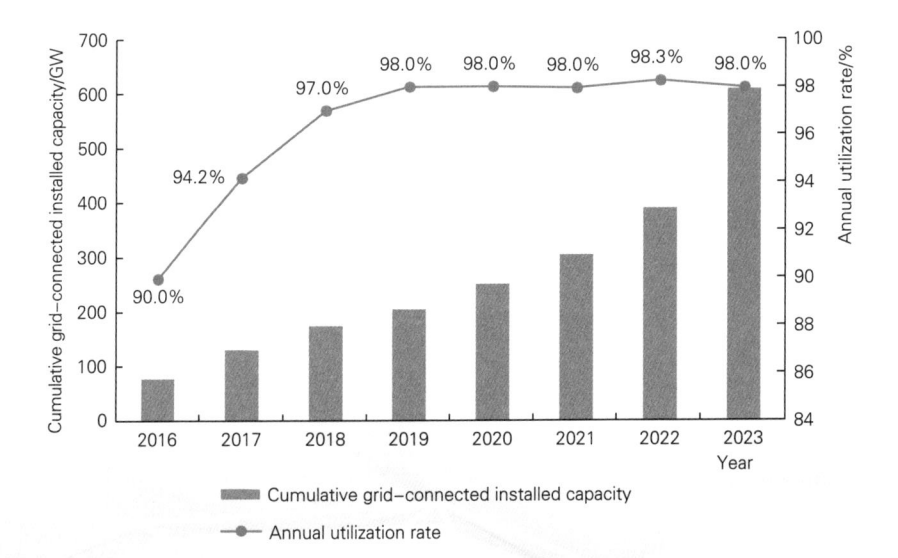

Figure 3.3　Cumulative grid-connected installed capacity and annual utilization rate of solar PV power in China, 2016 – 2023

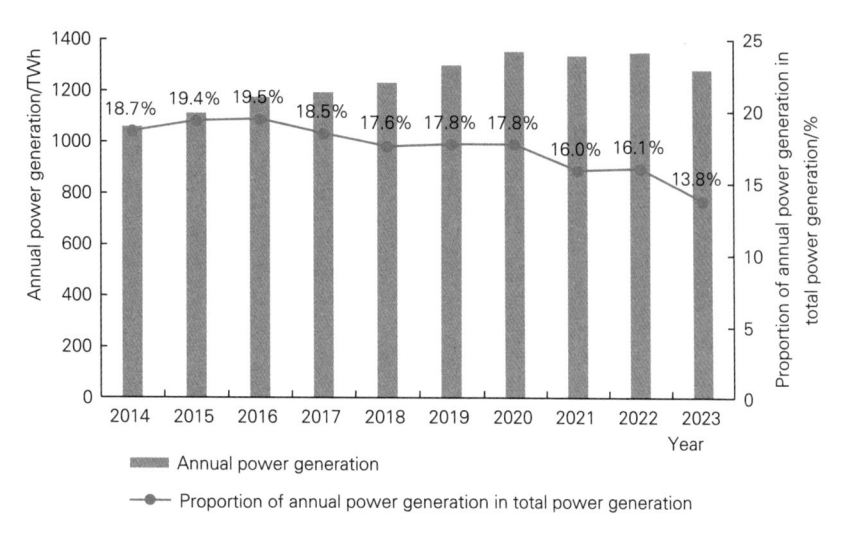

Figure 3. 1 Annual hydropower generation and its
share in China, 2014 – 2023

The share of wind power generation in China's power supply continued to increase, and the annual utilization rate reached 97. 3%

In 2023, the wind power generation in China reached 885. 8 TWh, accounting for 9. 5% of the total power generation in China. Coastal provinces actively developed offshore wind power projects to promote green energy transition. The average annual utilization hours of wind power plants in China amounted to 2,225 h, and the annual utilization rate of wind power reached 97. 3% (Figure 3. 2) .

The installed capacity of solar PV power plants increased rapidly while the utilization rate remained high, achieving an annual average utilization rate of 98%

By the end of 2023, the solar power generation in China reached 583. 3 TWh, accounting for 6. 3% of the total power generation in China. The solar PV power generation was 582. 3 TWh, of which centralized PV power generation was 359. 1 TWh; the distributed PV power generation was 223. 2 TWh, and it developed rapidly in North China, East China and Central China, with their power generation accounting for 87. 1% of the total distributed solar PV generation. The CSP power generation was 1 TWh. In 2023, solar PV power generation realized an annual average utilization rate of 98%, remaining the same as that in 2022 (Figure 3. 3) .

3 Utilization

The generation and utilization efficiency of renewable electricity continued to improve

The power generation of hydropower in China in 2023 was 1,283.6 TWh, and the annual average utilization hours of hydropower plants were 3,133 h

In 2023, the power generation of hydropower in China was 1,283.6 TWh, accounting for 13.8% of the total power generation (Figure 3.1). The annual average utilization hours of hydropower plants in China reached 3,133 h, and the effective utilization rate of hydropower in China was 99.0%, reaching a record high since 2016.

spring physiotherapy and agricultural breeding, and a small amount of geothermal energy was used for power generation. The large-scale development of shallow geothermal energy heating projects was conducted in North China and the middle and lower reaches of the Changjiang River, and middle and deep geothermal heating projects in northern China expanded in Hebei, Henan and other four provinces (municipalities).

The development of hydrogen production from renewable energy has been accelerated in China

China has been leading the world in hydrogen production and demand, with the total output of 42.91 Mt in 2023. The renewable hydrogen production capacity is 78,000 t/year, the capacity of projects under construction is about 800,000 t/year, and the capacity of projects in the planning stage exceeds 6 Mt/year. The production was mainly realized in the "Three-North" areas in China.

China has made a breakthrough in marine energy development

In 2023, China's marine energy industry made a major breakthrough, moving from theoretical research and small-scale experiments to large-scale engineering prototype demonstration: the world's first megawatt-scale floating wave energy generation device "Nankun" was put into trial operation in Zhuhai, Guangdong Province. The construction of China's first oscillating water column wave power converter equipment "Huaqing" with a capacity exceeding 100 kW commenced, and the 20 kW offshore floating thermoelectric power generator completed the sea trial in the South China Sea.

accounting for 20.9% of China's total installed capacity, becoming the second largest power source of the country. In 2023, the newly added capacity of solar PV power was 216.9 GW, accounting for 62.5% of the global total. The cumulative installed capacity of solar PV power was 608.9 GW, ranking first in the world. The cumulative installed capacity of centralized solar PV power was 355.5 GW with 121.1 GW added in 2023, and the cumulative installed capacity of distributed solar PV power was 253.4 GW with an increase of 95.8 GW. The grid-connected installed capacity of concentrated solar power was 0.57 GW.

The installed capacity of biomass power continued to increase; power generation from waste incineration entered into a stable development period; and the agricultural and forestry biomass power generation projects remained stable

By the end of 2023, China's biomass power generation utilization was equivalent to approximately 59.4 Mt standard coal, accounting for 73.9% of the total amount of biomass developed and utilized. The cumulative grid-connected installed capacity of biomass power was 44.1 GW with a year-on-year increase of 6.8%. The newly installed biomass capacity was 2.8 GW, of which the 0.6 GW was newly added to agricultural and forestry biomass power plants, accounting for 23%; the newly installed capacity of domestic waste incineration power plants was 1.9 GW, accounting for 68%; the newly installed capacity of biogas power was 0.3 GW, accounting for 9%.

The cumulative installed capacity of new energy storage in China exceeded 31 GW, with the energy storage capacity of lithium-ion batteries accounting for 97.4%

By the end of 2023, the cumulative installed capacity of new energy storage in China exceeded 31 GW, with an increase of nearly 10 times in the past three years, and the installed capacity in North China and Northwest China accounted for more than 50%. The energy storage capacity of lithium-ion batteries accounted for 97.4%, and compressed air energy storage entered the demonstration and promotion stage.

The large-scale development pattern of geothermal energy has taken initial shape in China, and the geothermal development was advanced gradually

Geothermal energy in China was mainly used for heating (cooling), followed by hot

construction was about 30. 64 GW. The installed capacity of newly approved large-scale conventional hydropower plants was about 4. 15 GW, estimated to generate about 18. 6 TWh each year. In the future, hydropower development will be expanded to high-altitude areas in the upper reaches.

The pumped storage capacity in China reached 51 GW, and the capacity of projects approved and under construction reached 179 GW

By the end of 2023, the total installed capacity of pumped storage reached 51 GW in China. The key areas for development were in East China, North China and South China, and the pumped storage was installed for the first time in Northwest China. In 2023, the newly installed capacity was 5. 15 GW, and the overall capacity of approved projects and projects under construction was 179 GW, with that of Central China, East China and Northwest China all surpassing 30 GW.

The installed capacity of wind power in China maintained rapid growth, with more concentrated development in large-scale wind power bases for onshore wind, and cluster development for offshore wind, as well as increased scale of wind turbines

By the end of 2023, the total installed capacity of wind power in China reached 441. 3 GW, accounting for 15. 1% of the total installed capacity in China and 43% of the global total installed capacity of wind power. The cumulative installed capacity of onshore wind power was 404 GW, in which the "Three-North" areas in China (i. e. Northwest China, North China and Northeast China) accounted for 66%. The cumulative grid-connected installed capacity of offshore wind power was 37. 3 GW, ranking first in the world for three consecutive years. In 2023, the newly installed capacity of wind power in China was 75. 6 GW, reaching a record high, with 69. 3 GW from onshore wind power and 6. 3 GW from offshore wind power.

The newly installed capacity of solar power in China reached a record high, with the accelerated construction of centralized solar PV, supported by the "desert and Gobi" base project, the steady development of distributed solar PV, and the rapid expansion of large-scale development of concentrated solar power (CSP)

By the end of 2023, the installed capacity of solar power in China reached 609. 5 GW,

China Renewable Energy Development Report 2023 Abridged Version 07

2 Development progress

The development of renewable energy has been accelerated comprehensively

The installed capacity of conventional hydropower in China reached 370.6 GW, with development advancing towards high-altitude areas

By the end of 2023, the installed capacity of conventional hydropower in China reached 370.6 GW, mainly located in Southwest China, Central China, South China, East China and Northwest China. In 2023, the newly installed capacity of conventional hydropower in China was 2.43 GW, and the capacity of large-scale conventional hydropower plants under

countries and regions around the world, reducing global costs. In terms of technological innovation, China has achieved remarkable progress in heterojunction cells, perovskite-silicon tandem PV cells, etc. , and has made breakthroughs in energy storage technology and smart grid technology. In 2023, China's investment in energy transition reached USD 676 billion[1], equivalent to 38% of the global total, ranking first in the world. These investments have improved the energy transition both domestically and abroad, supporting global clean energy projects and the construction of a sustainable and environment-friendly energy system.

---

[1] Source: BloombergNEF （BNEF）, *Energy Transition Investment Trends* 2024.

biomass energy utilization for non-electricity purposes was equivalent to about 20. 98 Mt standard coal, with a year-on-year increase of 24.4%. The annual production of bio-natural gas, solid biomass fuels, fuel ethanol and biodiesel has seen a significant increase. The large-scale geothermal development has taken initial shape to support mainly heating (cooling). The large-scale development of shallow geothermal energy heating projects was mainly conducted in North China and the middle and lower reaches of the Changjiang River, while the expansion of middle and deep geothermal energy heating projects benefited from the promotion of clean winter heating policies in northern China.

China's renewable energy policy system continued to advance

China has issued many policies in respect of legalization, marketization, technological innovation, carbon emission reduction, Green Electricity Certificate system and the establishment of standard system, so as to promote the healthy development of renewable energy. *Land and Space Planning Law*, *Energy Law* and *Renewable Energy Law* (Revised) were listed as major legislative items. Market-based operation has been standardized through documents such as *Basic Rules of Electricity Spot Market* (Trial), and market-oriented transactions of new energy have expanded. In terms of technological innovation, *Guiding Opinions on Promoting the Development of Energy Electronics Industry* was issued to explore smart energy systems. In the field of carbon emission reduction, China has boosted voluntary emission reduction of greenhouse gases and issued relevant administrative measures and project methodology. The Green Electricity Certificate system has been strengthened through *Notice on Full Coverage of Renewable Energy Green Electricity Certificates to Promote Renewable Energy Electricity Consumption*. *Guidelines on Standard System Construction of Carbon Peak and Carbon Neutrality* provided standards for wind power generation, solar PV power generation and other fields. Meanwhile, policies have also been promulgated for solar PV land use, concentrated solar power, wind farm renovation, etc. to promote the healthy development of the new energy industry.

China's renewable energy industry has become an important driving force for global clean energy transition

While promoting domestic clean energy development, China has actively provided technologies and solutions to facilitate global energy transition. China's domestic renewable energy industry provided high-quality products and services for more than 200

storage hydropower, 75.66 GW for wind power, 216.88 GW for solar PV power, and 2.82 GW for biomass power.

China's renewable power generation reached 2,951 TWh

In 2023, China's renewable power generation was 2,951 TWh, accounting for 32% of the total electricity consumption. The power generation of hydro, wind, solar and biomass was 1,284 TWh, 886 TWh, 583 TWh and 198 TWh respectively, accounting for 13.8%, 9.5%, 6.3% and 2.1% of the total power generation in China respectively (Figure 1.2). Wind and solar PV power generation accounted for more than 15% of the total electricity consumption, a year-on-year growth of 24%, becoming the main driver of the increasing share of non-fossil fuel energy consumption.

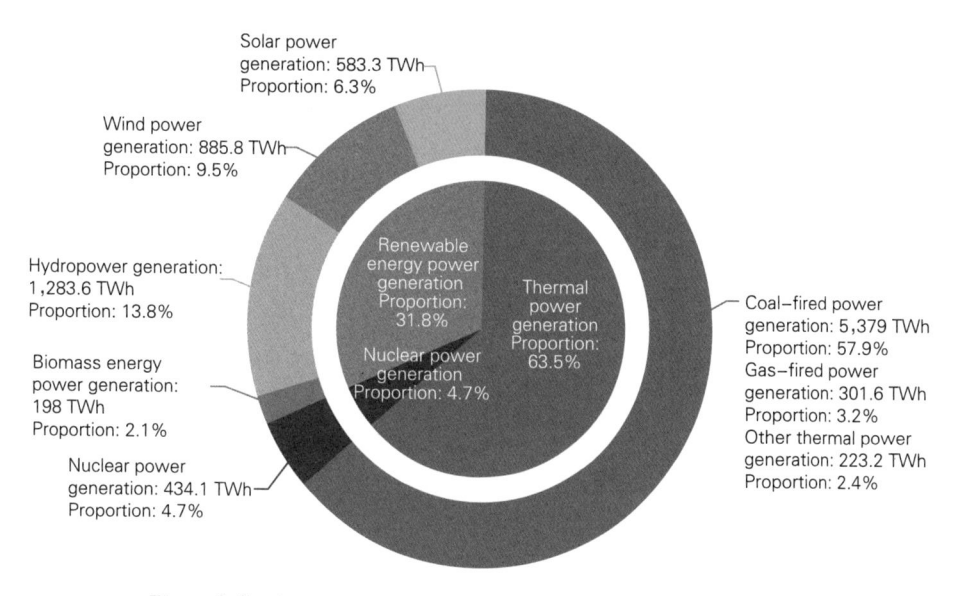

Figure 1.2 Annual power generation and its respective proportion of various power sources in China, 2023

Non-electricity utilization of biomass energy and utilization of other renewables in China have been steadily advanced

In 2023, the scale of non-electricity utilization of biomass energy and utilization of other renewables such as geothermal energy in China continued to expand. The amount of

China Renewable Energy Development Report 2023 Abridged Version　03

capacity（accounting for 12.7%）, 50.9 GW of pumped storage hydropower capacity （accounting for 1.7%）, 441.3 GW of wind power capacity（accounting for 15.1%）, 609.5 GW of solar power capacity（accounting for 20.9%）and 44.1 GW of biomass power capacity（accounting for 1.5%）. The total installed capacity of solar and wind power exceeded 1,000 GW, making them the second and third largest power sources in China in terms of installed capacity （Figure 1.1）. Renewable energy has become increasingly important in China's energy mix, and the installed capacity of renewable energy per capita exceeded 1 kW.

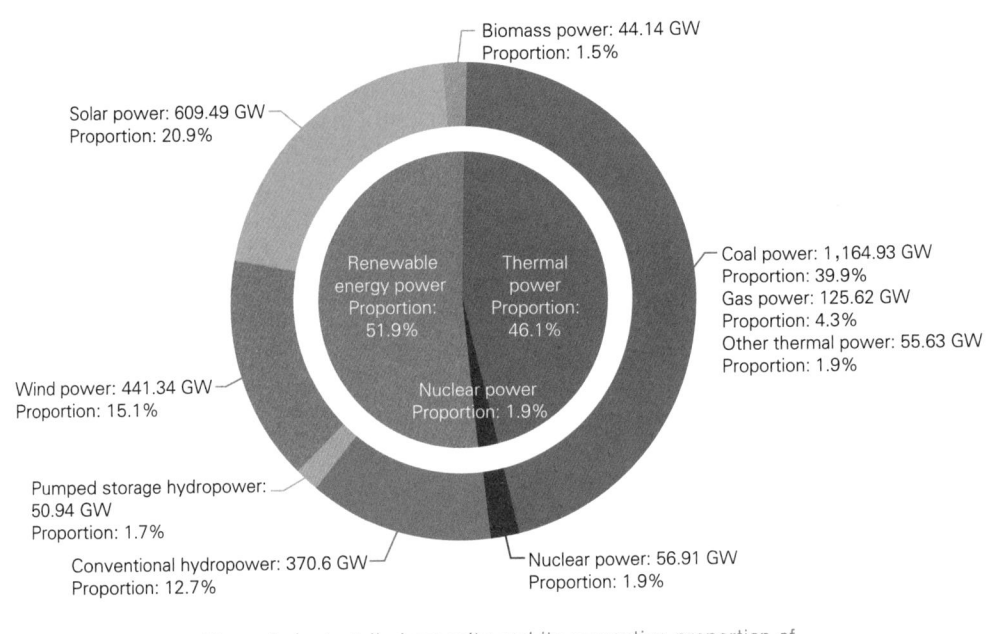

Figure 1.1 Installed capacity and its respective proportion of various power sources in China, 2023

The newly installed capacity of renewable energy in China was 303 GW in 2023

Since 2020, the annual newly installed capacity of renewable energy in China has exceeded 100 GW for three consecutive years. In 2023, the newly installed capacity of renewable energy in China reached 303 GW, exceeding the total renewable capacity in other countries around the world, accounting for 84.9% of the total newly-added installed capacity in China and 59.8% of the global renewable capacity additions. The newly installed capacity was 2.43 GW for conventional hydropower, 5.15 GW for pumped

1 Development status

China leads global renewable energy development

The cumulative installed capacity of renewable energy in China exceeded 1,500 GW

By the end of 2023, the installed power generation capacity of renewable energy in China reached 1,516.4 GW, with a year-on-year increase of 24.9%, accounting for 51.9% of the total installed power generation capacity in China and nearly 40% of the global installed renewable energy capacity. Specifically, 370.6 GW of conventional hydropower

CONTENTS

1 Development status: China leads global renewable energy development   01

2 Development progress: The development of renewable energy has been
accelerated comprehensively   06

3 Utilization: The generation and utilization efficiency of renewable electricity
continued to improve   10

4 Technological development: Continuous technological innovation promoted
industrial progress   14

5 Looking forward: China's renewable energy will become the key force
for a clean and low-carbon shared future   18

Editorial Board

Chairs:
Li Sheng
Yi Yuechun

Deputy Chairs:
WANG Zhongyao   HE Zhong   ZHAO Quansheng   YU Bo
WANG Fuqiang   Li Yangjia   GUO Jianxin   XUE Lianfang
QIAN Gangliang   Li Guangshun

Chief Editors:
ZHAO Zenghai   ZHANG Yiguo   PENG Caide

Associate Editors:
XIE Hongwen   GUO Yanheng   JIANG Hao   PENG Shuolun
XIN Songxu   ZHU Fangliang   XUE Jianfeng   YU Jianlin
Li Xiushu

Authors:
XU Zhicheng   XIE Yuetao   LIU Yuying   ZHOU Li
CHANG Haotian   XIAO Duanlong   SI Junlong   CHEN Zhang
DU Sijia   LIU Shuangxi   HUO Jingying   Li Shaoyan
MI Zhe   WU Mingxin   XU Yue   REN Weinan
ZHAO Duocang   WANG Huan   WANG Min   Li Yanjie
XUE Meijuan   ZHOU Xiaoxi   WANG Lingli   LYU Song
Li Xiaoxi   HAN Dong   AI Lin   CHEN Guosheng
KE Wen   CHENG Li   HU Yaofei   FU Zhaoqing
GENG Dazhou   JIA Haoshuai   YU Xiongfei   GUO Bao
LIU Chungao   MENG Xu   DUAN Cong   Li Hongyu
HAO Yu   ZHANG Bu'en   WANG Haoyi   ZHAO Guobin
GUO Jinyu   QIU Chen   QIAO Yong   CUI Zhenghui
XIE Hao   ZHENG Jing   WEI Guoqiang   ZONG Wanbo
QIU Yaijing   SUN Yirong   LIU Chao   WEI Huixiao
TIAN Zichan   HE Wei   HAO Jungang   ZHA Hao

CHINA RENEWABLE ENERGY DEVELOPMENT REPORT 2023

ABRIDGED VERSION